Shorebirds

A Compare and Contrast Book

by Sharon Dorsey

Kildeer with eggs in nest

Eurasian Dotterel

Shorebirds feed, breed, and nest along the shores of water. You might find them along the beach by the ocean, in marshes, by mountain lakes, or even in grasslands near rivers.

Red Phalarope

Western Sandpiper

American Golden-Plover

Common Snipe

American Woodcock

Shorebirds are master migrators. Many shorebirds breed in the Northern United States and Canada. They fly back and forth from their summer breeding grounds to their non-breeding homes on Caribbean islands, Central or South America.

When flying thousands of miles across continents and oceans, they stop at beaches, salt marshes, and grasslands to eat and rest.

Sanderlings

Short-billed Dowitchers

Dunlins

Many change their feathers (plumage) from one season to another. They do this to better blend in with their seasonal habitat.

Piping Plover migrating

Piping Plover female – summer breeding feathers

Piping Plover winter feathers

Shorebirds eat tiny fish, shellfish, little crabs, worms, and insects. Bill sizes and shapes help the birds get their favorite food.

Oystercatchers have bright red-yellow eyes and a bright, red-orange bill. Their long legs are perfect for wading in the water as they search for clams, mussels, and oysters to eat.

American Oystercatcher

The American Oystercatcher likes the sandy beaches of the Atlantic Ocean and the Gulf of Mexico in the US. They can be found on both the Atlantic and Pacific coasts of South America.

The Black Oystercatcher may be spotted on the rocky shores of the Pacific Ocean.

Plovers are small, round birds with strong bills.

Snowy Plovers are light brown with black or brown around their necks and short, black bills.

Piping Plovers have a sharp black collar, orange legs, and an orange and black bill. They are great at hiding on sandy beaches.

Snowy Plover

Piping Plover parent and hatchling

Semipalmated Plovers are brown with a thick, black band across their chest. Breeding adults have yellow legs and some orange on their bills.

Wilson's Plovers are bigger than Semipalmated Plovers. Their bills are also longer and thicker.

Sanderlings look like they are chasing waves but they are really looking for tiny snacks in the wet sand after the waves return to the ocean. They are medium-sized birds with black legs and bills.

Dunlins have a rusty back and a black belly patch. They are everywhere in the Arctic during the summer nesting season. You may see them in big groups near bays and coastlines during winter.

They have long, slightly-curved bills and eat little creatures just under the surface of the water.

The Marbled Godwit has a long, pointy orange-and-black bill like a sword. It uses this bill to dig deep into the sand and mud to find yummy insects and plant bits.

This bird is spotted brown cinnamon color, which you can see when it spreads its long wings to fly.

The Whimbrel is a beautiful brown bird with a long, down-curved bill. It summers in the tundra and can be found in mud flats, beaches, and salt marshes the rest of the year.

Ruddy Turnstones have orange legs. Their feathers are a mix of white, brown, and black. They have short bills that they use to look for food hiding under things on the beach.

Willets have brown feathers during breeding season and gray feathers in the winter. You can easily see them flying with a bold black-and-white stripe on their wings.

Red Knots have one of the longest migrations of any bird. They travel from the Arctic to the southern tip of South America and back each year.

Each spring, they stop on the shores of the Delaware Bay just as horseshoe crabs are laying their eggs in the sand. The knots eat the horseshoe crab eggs to get enough energy to continue their trip north.

These birds are chubby sandpipers with beautiful orange tummies and fancy gold, brownish-yellow, and black feathers on top in the summer.

A large family of birds called sandpipers (peeps) live near the ocean. The different species have different lengths of bills that they use to poke around in the sand and mud to find food. The different length bills allow them to live in the same area as they eat different foods.

The Least Sandpiper is tiny—about five inches tall, or the height of a soda can. They have cute yellow-green legs!

The Semipalmated Sandpiper is also small with a short neck and dark bill and legs. Its name comes from the little webs between its toes!

Least Sandpiper

Semipalmated Sandpiper

Spotted Sandpiper

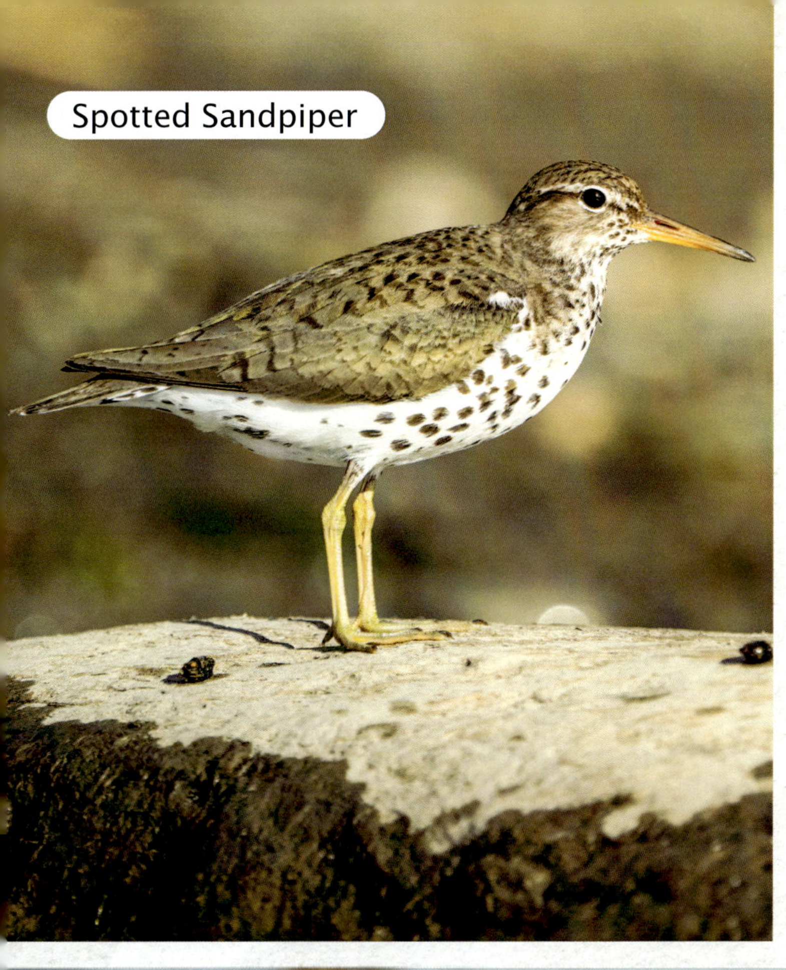

The Spotted Sandpiper is a medium-sized bird with brown spots on its white chest during the breeding season.

The Stilt Sandpiper is an elegant bird with long, yellow-green legs and a long bill. In the summer, it has a bright chestnut crown and spots on its brown-and-white feathers.

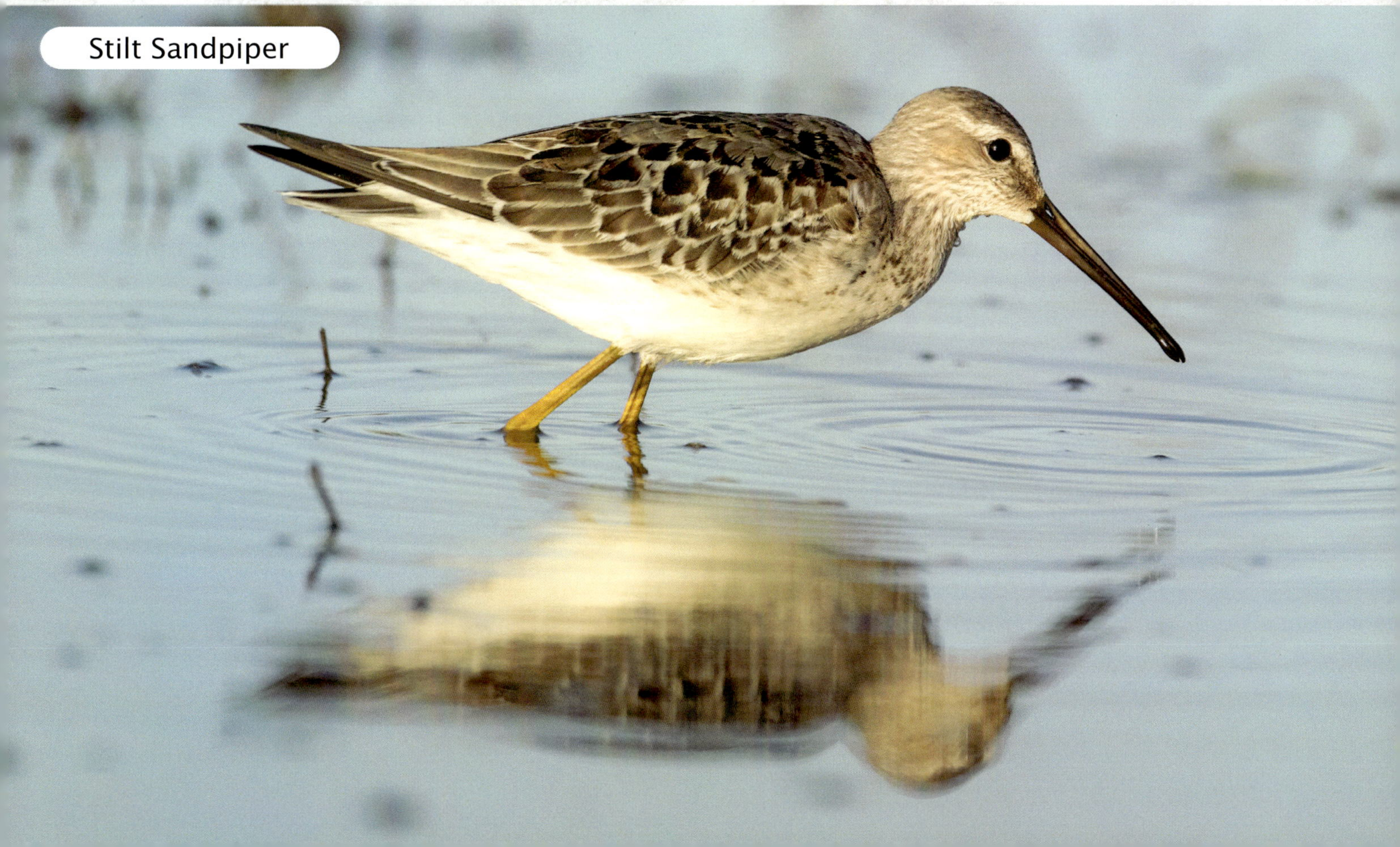
Stilt Sandpiper

Short-billed Dowitchers are colorful in the summer with orange, brown, and gold feathers. They have a special way of traveling called "molt migration," where they stop in different places to finish changing their feathers before arriving at their winter homes.

Long-billed Dowitchers look plain in the winter. Come summer, they have pretty black, reddish, brown, and gold feathers!

Greater Yellowlegs

There are two kinds of Yellowlegs. The Greater Yellowlegs is bigger and has a longer bill compared to the Lesser Yellowlegs.

Both use their bills to stir up shallow water looking for insects, small fish, marine worms, and snails to eat.

As you can see, shorebirds come in all different shapes, sizes, and colors! Sadly, these birds are having a tough time.

Migrating shorebirds often stop at the same rest areas (stopover sites) each year. If those areas have become overdeveloped or polluted, the birds may not be able to find food or a place to rest.

Some shorebirds nest right on the sandy or rocky shorelines. Humans and their pets may damage the nests without even knowing it.

Fortunately, there are some things YOU can do to help shorebirds. If you see them, don't chase them; please let them rest. If birds (any birds, not just shorebirds) nest on the ground, please watch where you walk and don't let dogs off leash where they might hurt the eggs or chicks. Last but not least, don't feed bread or crackers to wild birds.

For Creative Minds

Match the Adaptations

Match the body part adaptation description to the location on the bird.

1. My long, thin toes help me walk and balance in soft sand or mud. My long legs help to keep my body dry when standing in water.

2. My wings help me fly long distances. The colors and patterns of my feathers help me hide in my habitat.

3. I use my bill to dig in water and sand or mud looking for food. Shorebirds have different bill shapes to help them find the right food for each bird.

Long-billed Curlew

Answers: 1C; 2A; 3B

Mighty Migrations

Migrating birds use routes between their breeding grounds and their non-breeding homes that they know by instinct. They generally travel through one of three flyways: Atlantic, Midcontinental, or Pacific. When traveling, they will stop to eat and rest at the same beaches, marshes, and grasslands along the way.

Identify the locations the mighty migrators may visit each year.

> How far and for how long could you walk without stopping to eat or rest?

Red Knots (Atlantic Flyway) spend our winter (summer there) in Tierra del Fuego on the very southern tip of South America. They stop on a beach in Argentina and then again at the Delaware Bay to eat and rest. They breed in northern Canada and then fly back to Tierra del Fuego.

Buff-Breasted Sandpipers (Midcontinental Flyway) spend our winter (summer there) on natural grasslands of southern South America. They stop to eat and rest in Oklahoma on their way to their breeding grounds in the Arctic Circle, Canada.

Western Sandpipers (Pacific Flyway) spend our winter (summer there) on the Peruvian coast. They make several stops along the way before reaching breeding grounds in Alaska.

Shorebird Nesting

Most shorebirds are ground nesters. They scrape shallow indents in the sand, mud, or rocks to lay their eggs. The eggs are speckled to help to hide (camouflage) them from predators. Describe the different eggs and the nests.

Black Oystercatcher

Killdeer

Piping Plover

Sandpiper

Help or Hurt?

Which of these things do you think helps shorebirds and which do you think could hurt or harm them? Can you describe why? Can you think of ways you can help shorebirds?

1 over development

2 bird sanctuaries

3 pollution

4 marked nesting areas alerting people to avoid the area

Answers: 1: hurt; 2: help, 3: hurt, 4: help

A note from the editor: This book uses the names and capitalization recommended by the North American Classification and Nomenclature Committee of the American Ornithological Society.

Thanks to Arne J. Lesterhuis, Senior Shorebird Conservation Specialist, Manomet Conservation Sciences / Western Hemisphere Shorebird Reserve Network (WHSRN) for verifying the accuracy of the information in this book.

Library of Congress Cataloging-in-Publication Data

Names: Dorsey, Sharon, 1996- author.
Title: Shorebirds : a compare and contrast book / by Sharon Dorsey.
Description: Mt. Pleasant, S.C. : Arbordale Publishing, [2025] | Includes
    bibliographical references.
Identifiers: LCCN 2024033373 (print) | LCCN 2024033374 (ebook) | ISBN
    9781638173663 (trade paperback) | ISBN 9781638173670 | ISBN
    9781638173687 (epub) | ISBN 9781638173694 (pdf)
Subjects: LCSH: Shore birds--Juvenile literature. | Shore
    birds--Migration--Juvenile literature.
Classification: LCC QL678.5 .D67 2025 (print) | LCC QL678.5 (ebook) | DDC
    598.3/31568--dc23/eng/20240813
LC record available at https://lccn.loc.gov/2024033373
LC ebook record available at https://lccn.loc.gov/2024033374

English Lexile® Level: 970L

Bibliography:
"About Shorebirds." Western Hemisphere Shorebird Reserve Network. Whsrn.org, 2021, whsrn.org/about-
    shorebirds/.
AFSI | Cooperative Conservation for Shorebirds throughout the Atlantic Flyway. atlanticflywayshorebirds.org/.
"Cooperative Conservation for Shorebirds throughout the Atlantic Flyway". Atlantic Flyway Shorebird Initiative.
    atlanticflywayshorebirds.org/#x-section-8.
Discover Shorebirds Educator Guide L Curriculum L Learning Resources Grades 3-8 ACKNOWLEDGMENTS.
    https://whsrn.org/wp-content/uploads/2022/05/discover-shorebirds-050222.pdf
"Explore Bird Species | Bird Migration Explorer." Explorer.audubon.org, explorer.audubon.org/explore/species/.
"Habitats for Shorebirds." Manomet, www.manomet.org/project/habitats-for-shorebirds/.
"Home." Shorebird Flyways, shorebirdflyways.org/.
"Interactive Map – Migratory Shorebird Project." Migratoryshorebirdproject.org, migratoryshorebirdproject.org/
    explore-data/interactive-map/
"MBNMS: Seabirds and Shorebirds." Montereybay.noaa.gov, montereybay.noaa.gov/visitor/access/introbirds.html.
"Midcontinent Shorebird Conservation Initiative – Shorebird Habitat Conservation." Midamericasshorebirds.org,
    midamericasshorebirds.org/.
Pacific Flyway Shorebirds – Pacific Americas Shorebird Conservation Strategy. pacificflywayshorebirds.org/
"Search, All about Birds" Cornell Lab of Ornithology. www.allaboutbirds.org, www.allaboutbirds.org/guide/.
"Shorebirds 101: What to Look for When You Hit the Water." Audubon, 4 Aug. 2017, www.audubon.org/news/
    shorebirds-101-what-look-when-you-hit-water.
"Shorebirds | U.S. Fish & Wildlife Service." FWS.gov, www.fws.gov/library/collections/shorebirds.
Verlinde, Chris. "Share the Shore with Nesting Seabirds and Shorebirds!" Panhandle Outdoors, 1 May 2020,
    nwdistrict.ifas.ufl.edu/nat/2020/05/01/share-the-shore-with-nesting-seabirds-and-shorebirds/.